1995

W9-BPP-897

To
Suzanne,
When you need
a cosmic relief . . . . ,

Janice
Aug 95

E.T. 101

E.T. 101

The Cosmic Instruction Manual
for Planetary Evolution

An Emergency Remedial Earth Edition

ϟ

Co-Created by

MISSION CONTROL
and ZOEV JHO

HarperSanFrancisco
A Division of HarperCollins*Publishers*

HarperSanFrancisco and the author, in association with The Basic Foundation, a not-for-profit organization whose primary mission is reforestation, will facilitate the planting of two trees for every one tree used in the manufacture of this book.

⚡

FIRST EDITION

Library of Congress Cataloging-in-Publication Data
Jho, Zoev.
   E.T. 101 : the cosmic instruction manual for planetary evolution /
co-created by Mission Control and Zoev Jho. — An emergency
remedial earth ed., 1st ed.
      p.   cm.
    ISBN 0–06–251267–6 (acid-free paper)
    1. Civilization—Extraterrestrial influences.   I. Title.   II. Title:
ET 101.
CB156.J46   1994
001.9—dc20   95–1995
                                                        CIP

95 96 97 98 99 ❖ RRD(H) 10 9 8 7 6 5 4 3 2 1

This edition is printed on acid-free paper that meets the American National Standards Institute Z39.48 Standard.

DEDICATION

This third-dimensional publication of
*The Cosmic Instruction Manual
for Planetary Evolution*
is dedicated to all our relations.

—The Intergalactic Council

SHIFTS HAPPEN

Although it is not uncommon to revise a book in its subsequent printings, the author's name is not usually one of the items up for reconsideration. However, after the first printing of *E.T. 101,* the author's name has been mysteriously changed from "Diana Luppi" to "Zoev Jho," even though we admit to only slightly altering the text. What we do admit to changing extensively is the author herself.

In the course of bringing through this transmission, Diana Luppi's wiring experienced a meltdown. Her identity as an Earthling was seriously challenged, as was her once strongly held conviction that she was human. As a consequence of hanging out with us, Diana's identity went up in flames. What emerged from the ashes was a not even slightly human extraterrestrial master, known to us as Zoev Jho. Because that is all that remains from this incident, it is Zoev Jho's name that now rightfully appears on this manual.

Please be advised that the official Mission Control explanation for switching horses midstream is, "Shifts happen." In the course of reading this book, we caution you that a similar shift may also happen to you.

—*Mission Control*

THANK YOU

Mission Control thanks the following members of our ground crew and off-planet advisory committees for their assistance in the third-dimensional publication of this manual:

Extraterrestrial Earth Mission

The Ashtar Command

The Interuniversal Confederation of
Communication and Evolutionary Processes

The Telstar Pilot Project

⚡

Special thanks to

Council of the Empyrean

The Council of Twelve

Central Sun Productions

and

HarperCollinsSanFrancisco

TABLE OF CONTENTS

This is an Intergalactic Council Transmission

INTRODUCTORY INFORMATION

The Mission

(Reprinted by permission of the Intergalactic Council from the original Cosmic Instruction Manual, *Standard Interdimensional Pocket Edition.)*

The mission to Planet Earth was initiated by request of the planet herself. Earth has asked and the stellar councils have granted an evolutionary leap. Over the last millennium Earth has been preparing for our advent. It is now time to enter the Earth plane to reclaim this planet in the name of the Forces of Light and to open her doors to the cosmic community in which she resides.

We congratulate and salute those of you throughout the universes who have volunteered for this assignment. Go with our blessing. And remember, read the manual *before* you get there. Although you are veterans of countless successful missions to numerous dysfunctional life support systems, watch out for this one.

This edition of the manual is specifically designed for this planetary system—a system which defies all true rationality and has raised dysfunctionality to an art form. It is also one of the few systems where telling the truth is covertly considered a creative act. Because of the inherent hazards of this planet, Mission Control will not be responsible for any members of this mission who do not thoroughly acquaint themselves with the material contained in this special edition.

—*Mission Control*

Special Note from the Publisher

Although you were warned to acquaint yourselves with the original manual before departing, a great number of you did not. "If you've seen one manual, you've seen them all" has proved to be a cavalier attitude that many of you have lived to regret, even though most of you are too stunned and dazed by the process to remember just exactly what it is you are regretting. Being four hundred light years from home and suddenly wondering "Did I forget my toothbrush?" is both annoying to Mission Control as well as useless to the mission.

So, for those of you who left your galaxy without it, an unprecedented reprinting of the manual has been authorized by the mission's governing councils. This is the remedial version of the original manual because you couldn't possibly handle the unexpurgated version at this time. This edition is the official American and Canadian translation, written in the vernacular and made current to your circumstances. It is the strong recommendation of the Councils, now that you have yourselves totally disassembled, why not take a moment to read the instructions? It is, after all, at your request that we had them written.

—*The Intergalactic Council*

On behalf of the Confederated Interuniversal Councils, the United Stellar Alliances, and the Greater Interdimensional Federation of Light

A Word from Mission Control

Mission Control is the tactical arm of the extraterrestrial mission to Planet Earth. We implement the decisions of the stellar councils and act as an intermediary between the members of the mission who have opted for infantry duty on the planet's surface and those who are serving the mission in any of its many off-planet divisions. It is our responsibility to maintain all communication systems among the vast forces that have now gathered in this planetary arena. Our purpose is to assist the planet and its inhabitants into a new consciousness and reality.

Our prime directive is to coordinate the movements that Spirit orchestrates. We are here to assure that the decree of the High Spiritual Court is implemented and that the veils of the third dimension are parted so that light can enter. We have overseen the writing of this version of *The Cosmic Instruction Manual* at the request of the Intergalactic Council. We have done so because it is our mandate to assist all mission members in the successful completion of their many and varied assignments. Our instructions are to be a guiding force for this mission; our status is servant to Spirit.

This is Mission Control speaking. We now end this transmission.

How to Use the Manual

This manual is *not* a rule book, nor does it tell you how many angels are on the head of a pin and which dimension that pin is in. It also offers no street maps of the dimensions to satisfy your linear thinking minds. This manual is a tool to awaken you to what you already know. It is an aid in accessing the real inside information, not an outside authority with the latest "ism" in which you should believe.

Because we do not wish this manual to be misused, we have deliberately left out some information. If you find yourself wanting more answers than this manual offers, don't just stand there like a jerk; go inside and ask for those answers. Whatever is appropriate for you to know will be given. Mission Control would like you to be aware that we not only stand behind our product, but we also stand beside you as you read it.

We would also like to make it clear that although you may consider this manual a "new age" publication, we do not. That is because what you are calling "new age" is just the final manifestation of the old age. Since the "new age" is still about religious belief systems as asserted by outside authorities, we find it neither new nor much better than anything else you believe in. We prefer to think of ourselves as sacred cowpokes on a sacred cowherd, and our manual as a sacred cowpunching device. We believe that would make this publication more "post-new age" than anything.

How you end up classifying this manual is less important than how you end up using it. It is our suggestion that you use it as a tool to awaken yourself, not as a new doctrine to continue to snooze by.

This is Mission Control saying yeeeeeeeehaw!

1

The Intergalactic
New Collegiate Dictionary

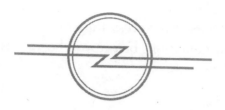

BECAUSE human languages are not designed to grasp many concepts outside the current fixated, fear-ridden, and toxic consciousness, and also because we do not use spoken language, this manual presents us with some interesting problems. Dealing in the currency of your linguistic systems is doubly difficult because the inversion of your energy has twisted all logical meaning. For instance, the people of this planet stand in total arrogance, adamantly denying their omnipresence. They declare their separation from themselves, each other, and all life while passing this off as an act of humility. Humility is not denial; *separation* is denial. And maintaining that separation is the ultimate act of pride.

Your institutions operate in the same backward manner; thus, you have a national security system that is actively engaged in killing everybody, a federal drug administration that has all but recommended motor oil for dietary use because it is low in polyunsaturates, and an economic system that has convinced everyone that life is bad for the economy. As you see, it may be difficult to use your language and still hope to express the truth. Certain terms need new definitions before you begin this reading. They follow.

⚡

Note: Some mission members have retained their extraterrestrial sense of meaning and may find this section a drag. Others have forgotten everything and require some reconstructive surgery. For that reason, we have highlighted the key definition—or as close as we can approximate a key definition in this language—with italics so that readers may either skim or read this section, as they wish.

Extraterrestrial

An extraterrestrial is not an alien. An alien is an alien. An extraterrestrial is a responsible citizen of the cosmos, not a foreigner adrift among the stars. Extraterrestrials are representatives of light, protectors of life, and lovers of the planets. They are indigenous to any planet they happen to be on by virtue of their citizenship, regardless of their planet of origin.

Many of you have come to believe that you couldn't possibly be an extraterrestrial because you feel so connected to the Earth and love her so much. May we suggest that if you love this planet at all, you *are* an extraterrestrial. May we also suggest that your concern for this planet was so great that you cared enough to send the very best—in this case, yourself.

Alignment, not lineage, defines the meaning of the term "extraterrestrial." Although all life emanates from the same source, not all life is aligned with that source. ***An extraterrestrial is a being who is in sympathetic harmony with the essence of its genesis.***

Alien

The people of this planet have expressed a certain neurotic fear about an alien invasion. That fear has been triggered by a nagging sense that in a limitless universe there just might be other intelligent life. In a typically xenophobic and self-serving response, governments are arming against what they already know to be true. This, of course, is not general knowledge but, in paternalistic systems of government, important information is *never* general knowledge. In the name of national security, the acts of government are often hidden from the governed. This statement is not to make you paranoid or stir you to political revolt; it is simply to demonstrate to you the behavior of an alien and the meaning of the term.

You need not scour the skies for evidence of an alien invasion. Look around you instead. Look at those who are peddling fear, vending death, and poisoning the planet. Look at those who hide the truth so that the power they have derived from lying will not be threatened. And look at those sadly separated beings that have the audacity to gaze out upon the created universe, wondering if there could possibly be anyone else out there, all the while arming just in case there is. This is the alien invasion you worry about, the one you externalize and fear. It is also the one that surrounds you and the one you have been living out for thousands of years.

There is no point in fearing an invasion of aliens, since the invasion has already happened and the aliens are already here. You would be better advised to fear that no true intelligence will ever show up on this planet. And you would be even better advised not to fear anything at all.

E.T. vs. Alien

Now that you have learned the difference between an extraterrestrial and an alien, we would like you to forget the distinction immediately. The danger in the definitions is that, as mental concepts, they separate once again. This mission is not about separation. Nor is it a Hollywood Western being performed by a cast of good guys and bad guys. It is about light and bringing more of it onto this planet. *The invitation to enter into the light is extended to all humankind, aliens included, for aliens are only extraterrestrials who have chosen to stand in the darkness, live a lie, and wear a disguise.*

Note: At the time of this printing, there are only two basic types of people on this planet: extraterrestrials and aliens. "Extraterrestrial" is a transitional term which will become unnecessary by the completion of this mission. At that point, *human* will replace the term. An awareness of your extraterrestrial nature will then be an integral part of all human experience, and aliens will no longer occupy the planet. In the same manner, other transitional terms, such as "androgyny," will cease to have any meaning. A balanced male and female will emerge within every being. As a result, the word "androgyny" will be thrown on the trash heap, relegated to the status of a needless term that redundantly describes what it means to be human. Keep in mind that the definitions in this dictionary were written for a world in transition and are subject to revision.

Walk-In

Most of you have probably heard of the term *walk-in,* but for the benefit of those who have been assigned to some real boondock outposts, we will explain its meaning. A walk-in is a member of the mission who has "walked" into a body that was previously occupied by another tenant. The main function of walk-ins is to assist ground crew members who came here in a more conventional manner to awaken to their true identity—hopefully before the mission is over. They retain much of their interdimensional consciousness and can move through dysfunctional patterns at an accelerated rate, making them invaluable to the numbed-out and befuddled crew members who have been here their entire lives. The walk-ins are an expeditionary unit, most of whom will walk right back out once the task of awakening this planet is completed. ***Walk-ins are emissaries of the light who are serving in this mission's rendition of the Foreign Exchange Program.***

Crawl-In

Even the most urbane and knowledgeable members of this mission will not have heard of the term *crawl-in* because we just made it up. ***Crawl-ins are planetary transition team members who opted to enter this plane through the normal, currently traumatic birth process.*** Upon arriving, most of them instantly reevaluated the situation and changed their minds, but were unable to figure out a way back.

The majority of this group incarnated shortly after World War II, their advent being triggered by the Manhattan Project's birthing of the nuclear age and the subsequent atomic bombing of Japan. They are referred to as the "war-baby crop" or "baby boomers" by the unsuspecting local population. The crawl-ins are the backbone of this mission. If you are

reading this, you are most likely one of them because the "crawl-ins" are the ones that necessitated
the writing of this manual.

UFO

Some people on this planet are certain that UFOs
are currently visiting this place. Most people feel
that this is an absurd allegation of a marginal group
whose members are basically nuts. This perception
is backed by governmental agencies that swear that
UFOs do not exist and have tons of highly classified
information to prove it. The UFO advocates justly
point out that no government needs secret files on
something that doesn't exist. They find it equally
absurd for governmental agencies to simultaneously refuse to release such files on the grounds
that national security is at stake. Most people
haven't given this slight inconsistency any thought,
going along with the party line under the assump-
tion that the government wouldn't lie and that "fa-
ther knows best."

Mission Control would like to put this matter to
rest. The UFO faction is made up of what we call
the "nuts-and-bolts" people. They are still intrigued
with third-dimensional phenomena and are consequently missing the boat while looking for ships.

Those who believe there is nothing to believe have been basically brainwashed by official discrediting of the question and are equally off track. The governmental agencies are lying through their teeth and are therefore the most off-base of all three factions.

Third-dimensional vehicles from other planets do, of course, exist. However, they are not the ships you should be preoccupied with. If an object can be identified as an unidentified flying object, it is not one of the craft that we control, nor is it a member of the Royal Celestial Air Force fleet that is the right arm of this mission.

Our craft are not third-dimensional; however, they are in position throughout your skies at this moment. They land wherever and whenever they wish, and they are not using an "invisibility cloak" to hide their presence. They are blatantly in the open, visible to the few who have broken through the blindness that afflicts the third dimension.

We do not mean to trivialize your third-dimensional experience or discredit the craft which travel that dimension. There are many very sweet entities in the third dimension who are assisting with this mission; however, their vehicles are definitely in the General Motors branch of this expedition and not representatives of the mighty force of fifth-dimensional craft that now gird your planet's atmosphere.

As you awaken, the presence of other dimensional craft will become obvious and fill you with awe. As a result, the contested issue of the 3-D UFO will fade in interest, just as the Model T Ford no longer thrills you or occupies your thoughts.

⚡

Note: The purpose of this piece is not to define UFO, but to clear up some of the planetary provincialism that surrounds this issue. It is also to help prepare you for the upcoming scandal of "Cosmicgate," a disclosure of the intergovernmental cover-up of extraterrestrial presence that is soon to be exposed globally. Because this entry serves another function and is not a true definition, no part of it warranted highlighting.

Note: For clarity, we now offer a brief description of the dimensions: The **third dimension** is the one you are currently living in and transiting out of. It is the one you consider the sum total of reality. The **fourth dimension** is sometimes referred to as the astral plane and exists as a shadow dimension to the third. Like the third, it is also a dimension of polarity and is inhabited by what you call "spirits" and disembodied entities. This dimension has fallen out of favor with the thinking of scientific materialism and has been reduced to the ranks of a primitive, superstitious belief—a belief that permeated human myth until you all smartened up and dismissed it. You may be surprised to learn that the truth does not require your belief in it in order to function, and the fourth dimension has managed to carry on despite your rejection. The **fifth dimension** appears in your symbol systems as "heaven," and, compared to the third dimension, it is. It is a dimension of light and love, and it is free of the illusions of duality and separation. The fifth dimension is in no way the end of the line; it is just the next step in your planetary evolution. Creation actually contains an infinite number of dimensions, many of which you inhabit simultaneously. We hope that clears this matter up for you.

LEVEL II WORDS
Look Jane, See Spot Transmute

Light

The concept "light" is a misunderstood term. Few have grasped its meaning, and most use it lightly. Since the manual uses this term often, it requires an expanded definition.

True light is awesome. It is so far beyond its common English usage in conjunction with such things as neon, stop, sun, flash, Bud, and "do you have a. . . ?" that it is difficult to express its actual meaning in this language. Let us put it this way: You are victims of indirect lighting. Whatever romance you have created by using light in this manner has lost its charm. Direct lighting is the wave of the future. As a mission member, you are specifically here to plug into that high-voltage line.

Please become conscious of this word's meaning as you use it. *Light is the force of reclamation, stewarded by the power of creation. Light is nothing less than life itself.*

Transmutation

Transmutation is not to be confused with transformation. This world had to go through thousands of years of transforming before it was in a position to transmute. That cycle of transformation is now complete, and the transmutative cycle has begun.

Transmutation is a genetic change at the cellular level, which is now in process for <u>all</u> life forms on this planet. The Earth, who is a living consciousness, has made her decision, determined her course, and begun her dimensional shift. Subsequently, all planetary life is being prepared for this event through the cellular transmutational process. This is a birth process which will deliver this planet and all participating life forms into the fifth dimension.

Cellular transmutation is not something you may choose to do if it interests you, like taking up golf. It is something that is happening and that you chose to do before you got here; otherwise you wouldn't be here. Although you have no genetic option in this matter, you still have free will. You can willingly assist this procedure and transmute with this planetary sphere, or you can resist the process and become, as some members of our planetary transition teams like to say, "crispy critters." Mission Control advises you to think twice before you change your mind.

Intelligence

We have noticed that your idea of intelligence and our idea of intelligence have very little to do with one another. For instance, you call yourselves an intelligent species, yet you are dangerously close to making your planet uninhabitable by anything other than asphalt. You have also managed to place yourselves at the top of the endangered-species list. May we point out that even a virus demonstrates a more astute grasp of its situation than that. The only reason a virus is inclined to trash out its environment is in its well-calculated attempt to maintain its life.

We have also noticed that you use the word "smart" in conjunction with business swindles and corrupt deals. When someone sells property that is located on a quicksand bog, you say, "Boy, was that a smart move!" You also think it is incredibly clever to sell a used car for top dollar without mentioning that it has no transmission. Both these examples are trumped-up illustrations that lack the malignancy of your actual activities. Your governments, your corporations, and your citizenry commit mind-boggling atrocities in the name of material cunning, and all human commerce is riddled with spiritual scandal. Moreover, though such acts may technically be fraud according to your laws, fraud is an issue only if you have the misfortune of getting caught. Otherwise, these acts remain shrewd business moves, the products of brilliant minds. For obvious reasons, we are perplexed by your concept of intelligence and would like to offer another definition.

The basic misunderstanding of this term arises from the fact that the inhabitants of this planet have confused brains for intelligence. A brain is an instrument of intelligence, while intelligence is a force. ***Intelligence is the force of life expressing itself in created form.*** It exists in all life, regardless of whether it has a brain or not. Through the misuse of your mental processes, you have come to regard intelligence as the art of one-upmanship in acts of spiritual barbarism. You have somehow managed to reduce rationality to the mental faculty that enables you to grab the most the fastest. Meanwhile, true intelligence is an alignment with the matrix of creation and its source. It is an allegiance to light and an embracing of life, not a demonstration of how adept you are at the act of denial.

Our purpose here on this planet is to assist in freeing you from denial so that you can finally begin to think straight. It is our mandate as well as our longing to help raise you out of your deranged thought process up to the status of a truly intelligent life form.

Co-creation

This concept may be difficult to get across because you neither acknowledge your part in "creation" nor do you know anything about "cooperation." Although both words appear in your dictionaries, we have yet to see either evolve beyond the status of a good idea.

Please do not mistake your empire-building for creation or your mutual palm-greasing for cooperation. Creation is not about razing a site and then rebuilding on that recently destroyed location, nor is it demonstrated by erecting prison systems and euphemistically calling them cities. One reason the true meaning of creation escapes you is that it cannot be accomplished by people who see themselves as victims. That disqualifies most of Earth's population right there. Another reason is that it is founded on cooperation, which eliminates all those who were not eliminated on the first count and brings us to the other word you do not understand.

If Earth's inhabitants had any idea what cooperation meant, the longest distance between two points would not be a committee. Nor would countries march endlessly from one war to the next. Your current understanding of cooperation is based on concession traded off for gain. That is why an institution like the United Nations is ineffectual. It has neither the bargaining chips nor the tools with which to play this global game. As long as the inhabitants of this planet remain powerless in their reality and separate in their intention and self-definition, co-creation will remain a pleasant, though unworkable, concept.

The reason we are trying to explain a word you are ill-equipped to understand is because it is the basis of this mission, and its entire plan. We are here to aid you in acknowledging your power. We are here to help you take command. We stand at your threshold, calling you to courageously walk away from the lie of human separation into the arms of your new and ecstatic lives. Once you have begun to take that first step, we will walk as your allies at your side. We will then be able to extend you our services, joyfully assisting you in the creation of your dawning, light-filled world. At that point you will finally understand that *co-creation is the awesome life-giving, light-bearing act of equals, powerfully operating out of an integrity that is aligned with the truth.*

Unfortunately, that brings us to another word you do not comprehend. Please go on to the next definition if you are curious what "truth" means.

Truth

We have noticed that you pretend to value truth on this planet. Some spend a lifetime seeking it. Your legal systems demand it, and you can be sued if your business doesn't practice it. Your philosophers define it, your scientists measure it, your religions exalt it, and you all fight over it. Meanwhile, all you are doing is paying global lip service to it. There is an excellent reason for all this: You have no idea what truth really is.

How the obvious has escaped you is a tedious story. The abridged version of it amounts to this: You embraced fear. After that unholy act, it has been downhill ever since. Fear is the first lie, the lie that tells you that you are separated from the whole. Once it has been embraced, you are incapable of ever telling the truth under any circumstances without blowing the game.

Truth, by its nature, is the light. Fear cannot, by its nature, be in the light without dying. It becomes a simple matter of self interest. Fear has owned this planet, its people, and their systems for a long time. It does not wish to give up the property it has acquired because it is a parasitic life form that cannot live separated from your life forces.

The truth is, you are the truth. It is not external to you, as you have been led to believe. For that reason, it is ludicrous to set out on a spiritual journey in search of it. It is likewise ridiculous to punish those who do not practice it when almost nobody on this planet does. As for philosophizing over it, how can you when you wouldn't recognize it if it ran you over in the street? Meanwhile, measuring it is done in your attempt to dominate it, leading you further into the lie that it lives outside of you like an enemy that must be controlled. To exalt it is also to see it as separate. And fighting over it is so absurd as to not deserve our comment at all.

The totality of your clinically insane behavior surrounding truth has been cleverly manipulated by fear in its attempt to keep your eyes *off* the truth. In this manner, fear was able to continue uninterrupted and undetected in its process of eating you alive. But don't worry—there is a cure. All you need do is awaken to the fact that *you are the truth*. As the light comes on, the parasite will die, leaving you joyously able to reclaim command.

Reality

This is a difficult word to define because there really isn't any such thing. What we mean by that is that there is no single reality, here or anywhere else. There are as many different realities on this planet as there are people alive to create them all. And

what passes for global reality is merely a group con-
sensus on a few minor points. From there on out, it's
every man and woman for him or herself.

*The reality that you live is nothing more
than an audio-visual demonstration of where
your attention is.* The universe presumes your attention is on what you want and graciously provides
you with more of the same. If this dynamic were
understood, you would never do anything so foolish
as to declare a war on drugs—unless, of course,
your objective were to create more of them. There
is no better way to increase drug traffic than to have
everyone's attention focused on it. This same princi-
ple applies to increasing everything you think you
oppose, and it is also the reason a war cannot be
won. If you were serious about stopping drugs, the
best course would be to stop being fascinated with
them via your perceived opposition. Become fascinated with freedom instead, and your addictions
will disappear naturally to satisfy your new preoccu-
pation.

Because you have yet to understand your power
of creation and who you really are, you perpetually
put your attention on denial instead of affirmation.
This results in the universe serving up an extra help-
ing of what you thought you didn't want. Although a
number of you practice the art of affirmation as a

tool for changing your realities, you can affirm until you are blue in the face and they may fail. Unless your attention and your identity have also been altered to accommodate what you affirm, the universe has no option but to fulfill your real, though hidden, attention's desires. Until you understand the role that your attention and sense of identity play in your creation, your affirmation track record will remain a perplexing hit-and-miss affair.

It is time to use your powers wisely and create realities that are worthy of who you are. You can do this by changing what you communicate to be real to the universe through your focus, and the identity you project by way of that identity's behavior. If you do not make this fundamental shift, you will continue to transmit the same old tired requests to a universe that will dispassionately and lovingly respond with the same old tired and often toxic answers.

Spirit

This is the most important word in this manual. Spirit is the driving force behind this mission as well as its designer. It is the reason our great forces are now assembled here. It is also the reason you, our mission members, willingly chose to incarnate on this seemingly hostile and backward planet. We all came at Spirit's call.

Everything and everyone in every universe rightfully belongs to Spirit. *Spirit is the power that breathes life into all created form. Spirit is life's true identity, as well as its long-awaited beloved. Spirit is the source, it is love, it is all.* And, although Spirit has always resided here, it has now chosen to lift the veils that have kept its presence on this plane from being fully known. It is out of love of Spirit that the vast Forces of Light are now infiltrating every earthly system. We were summoned here to participate in the transfiguration of this planet into the glorified home of Spirit it is destined to be.

⚡

We could go on redefining your language forever because, at the moment, there is very little agreement between us, except on words like dog and cat (and even there the agreement is minor). Since philology is not the primary intent of this manual or this mission, we will leave the matter of meaning at this point, knowing that your language is about to change organically as the natural consequence of the imminent change in your consciousness.

2

Transmutational Procedure

RULES FOR
DYSFUNCTIONAL PATTERNS

Step I:
In Rome, Do as the Romans

Upon arrival on the Earth plane, your instructions were to completely fall asleep—just like the local population. You were to totally forget your true identity and everything you knew.

Since most of you entered as babies, this was not difficult. Every institution in the culture supported this memory loss, and it became easier as the years went on. Any inadvertent slips on your part most likely occurred during childhood and were easily dismissed as the result of an overactive imagination.

Since imagination threatens the dysfunctionality of this world, it was probably drummed out of you as soon as possible by the adult inhabitants of the planet. What your parents were unable to suppress, the school systems most likely made short work of, as this is their specialty. In this manner, the local planetary inhabitants unwittingly assisted in maintaining the secrecy of your presence and the security of the mission.

Step II:
In Rome, Do as the Arcturians

Ground Rules

Step II of the transmutational process cannot begin until the successful completion of Step I. In short, you must be able to pass for a local, and you are not allowed to just fake it. Total dysfunctionality must be achieved before Step II can commence.

When extraterrestrial incarnates on mission to Planet Earth finally arrive at the point where they are no longer able to demand water they can drink, food they can eat, or air they can breathe without killing themselves, they are to understand that Step I of the mission has been successfully completed. The incarnates have truly become Earthlings, and Step II may now begin.

Coming Out of the Closet

You may dimly recall the saying, "In Rome, do as the Arcturians." If not, don't strain your memory. Even if you do remember, the humor of it may not be immediately evident. That catchy little intergalactic saying was coined to capture the essence of Step II of the transmutational procedure. That procedure entails waking up to your true identity and forgetting everything you learned up until this point so that you can remember what you actually knew before you got here. In other words, you are to junk the entire identity you just spent a lifetime laboriously creating. Now do you see why we say the humor may escape you?

All Roads Lead Away from Rome

Yes, you understood the preceding entry correctly. You are to disengage yourself from your old identity and disassociate from a declining Rome. After falling asleep profoundly, you are now expected to wake up, equally profoundly. Now is the time to dismantle all false identity. Now is the time to forget that which has been learned in deference to that which is deeply known. Now begins the awesome process of altering human history. Now is the time for everything, and now is here.

⚡

(Refer to "Time/Space Anomalies and Their Physical Functions" in this section for further clarification on "now." Also see "Passing for White" and "Closet Cases" in *Troubleshooting* for some precautionary instructions about this emerging consciousness.)

Gentle Reminder

Some of you are probably wondering why such a torturous route was chosen to get to the desired destination. The reason you are wondering this is because you have been on this planet too long and have absorbed some, if not all, of its dysfunctional thinking. Keep in mind that this planet is no model for rational thought, and that what passes for sanity here is sending chills down the spine of the remainder of the universe.

The need to absorb the dysfunctionality of the planet is in order to legitimately disarm its patterns. Any other method would constitute an invasion, *and we do not invade.* We alter by earning the right to do so. No entity is permitted to enter an alien world and disarm its dysfunctional patterns without having lived them. That is in compliance with Universal Law, which we represent.

Although we have had transmissions from many of you, screaming, "Invade already. Just get me out of here!," we regretfully remind you that that is not what you signed up for. Getting out of here is not the point. Getting more light *into* here is. Remember?

SOME INTERESTING FACTS ABOUT TRANSMUTATION

Pre-encoded Activation

The seeded entities that comprise this mission were pre-encoded to awaken at this time. That means that their DNA structures were genetically designed to go off, like time bombs, at a designated moment in order to accommodate more light than the current human model was prepared to entertain. The time for genetic detonation is now.

Time/Space Anomalies and Their Physical Functions

A. *The Dimensional Shift:* Time/space anomalies are being experienced on your planet at this moment. Most people have the nagging sense that there isn't as much time as there used to be. This is usually expressed as, "My, isn't time flying!" The reason for this sensation of less time is because there *is* less time.

In order for a dimensional shift to occur, time is collapsing to create a new dimensional space. Conversely, space is collapsing to create a new dimensional time. In other words, the time/space relationship that determines your third-dimensional reality is up for grabs.

"Now" is an actual event. It does not refer simply to living in the present, although that is strongly recommended. A time warp is truly occurring and will continue to accelerate until "now" is fully reached, and the dimensions can finally interface, much like a ship docking in a spaceport at a predetermined moment.

Now there is more "now" than there was even a few months ago, and even more "now" is on the way. When Mission Control or any of its alliance members state "the time is now," we are reminding you of your genetic agreement as well as being literal.

B. The Genetic Shift: Another interesting aspect of this anomaly is that your DNA codes were set to go off "now" before you left. As you may have noticed, that did not occur at your birth, or at any other time, until now. Even though you have lived through many "nows," you have not yet lived through this one. The genetic shift is triggered by "now" to accommodate more "now," and the degree of your awakening is in direct relation to the degree of "now" that you are experiencing because "now" determines that awakening.

Note: In an emergency situation, "now" can be brought about instantaneously and all DNA codes activated simultaneously. This is a bit like calling up the reserves, and we would rather not use this method if possible. A more organic approach is preferred. (Please see "Emergency Procedure" in *Troubleshooting* for more information.)

Deprogramming

As the transmutation of genetic structure unfolds throughout the planet, your pre-encoded genetic program will activate. For a time, you will groggily begin the process of deprogramming from the old systems. Your identity will begin its march out of the third dimension while your personality and ego may be more inclined to cling to a sinking ship.

As the old programs of a dying world begin to unravel, you may experience a little discomfort, such as your entire world falling apart. It may be useful during this transition to remember that you are an interdimensional master who is an expert at transmuting crumbling realities. You have done this many times before.

Debriefing

All that you have endured during your residency on Planet Earth is extremely valuable to Mission Control. What interests us is not the information concerning the nature and effect of human dysfunctionality. If that were all we wanted, we could have just as easily read a newspaper. However, the fact that you endured human denial *is* of great importance. Why? Because, as you transmute, all the dysfunctional patterns that you have willingly taken on will transmute with you.

Although you may not yet recall, you agreed to do this to assist the Earth in her birthing process into the light. The nature of the agreement was that you would willingly transmute the denial you have borne, who you have been erroneously told you are, and all that you have come to humanly represent. You agreed to transmute all this into the very fabric that is to become the new garment of a transfigured world. Consequently, your debriefing process is very dear to us because it is a sacred act.

Helpful Hints for the Second Coming

The Second Coming is imminent, and you may as well get ready. This is a particularly good idea because you're it. You *are* the Second Coming. Mission Control does not wish to stay on this topic very long because we are aware of the charge that surrounds it due to 2,000 years of organized denial. For this reason, we will give you only one more helpful hint: Become your own Messiah—why wait?

That statement is not only *not* heretical, it is the entire point. That is what the transmutational process is all about. Mission Control has no further comment.

3

The Mission

THE FOLLOWING section uses many military terms to describe the activities of the various branches of this mission. It is important that you remember we are not engaged in military maneuvers in any sense that the people of this planet understand those terms. We are not here to force a change because we know that force changes nothing, and we are far too interested in change to even try it. We are not at war, and there are no bad guys to defeat—just an offer of loving assistance to the people we came to liberate.

Your belief in the enemy's existence will soon pass as your planet transits out of the third dimension where that illusion resides. We did not choose military terminology to foster your fictional belief in an external adversary. We only chose these terms because we had to choose something, and you all seem very conversant and comfortable with military thinking. Your focus was the basis for the analogy, coupled with your need for mental structure. This terminology is in no way an adequate expression of who we are or what we are doing. Unfortunately, any other analogy we could have chosen would have been equally inadequate because it is virtually impossible to package a fifth-dimensional thought form in a 3-D wrapper.

If you do not like our use of your language, please keep in mind that that is exactly what it is—*your* language. That is why we would have preferred that you had acquainted yourselves with this manual before your arrival, on worlds where communication is pure and where misinterpretation is impossible. Be aware that until you are fully reawakened, there is always a potential for misunderstanding. Our advice is to proceed with caution and to not mistake the mental concepts we present as expressing the full meaning of what we say. Also, keep in mind that this mission is not a military threat. Its sole purpose is to assist you and your planet through a graceful transition into the light, and its sole motivation is love.

Job Titles—An Overview

When the process of awakening is completed—and even as it is unfolding—you must begin to take your posts. For each of you, that post is different. It is beyond the scope of this manual to list all the job descriptions that you are soon to fill because each of your tasks is unique and designed around your essence. Therefore, the knowledge of your innate purpose must come from within. You will know when you are doing your specific work when there is a deep resonance within your being. The ease and grace with which your life flows is also a clear indication that you are aligned with your true function.

What follows are broad areas of the mission's directives. Some topics are being left out deliberately because of their highly sensitive nature. The material contained in this section of the manual will be enough to trigger your memory of those areas that are not mentioned. Be aware that this is a team effort; since no two positions are identical, each one is indispensable. Mission Control is depending on all of you, your mastery, and your thorough training to accomplish the tasks at hand. May the Force be with you.

The PLO—Definition and Purpose

Regardless of your specific tasks, you are all members of the PLO (Planetary Liberation Organization). This is a spiritual organization and is not to be confused with the political, Earth-based PLO that is constantly in the newspapers. That group, like most human political movements, is fear-based and hate-driven and, as such, has nothing to do with us or what we are doing.

Do not mistake this statement for Mission Control's stance on the Near East. Mission Control has no political stance on anything. We have a spiritual stance that views your PLO with the same weariness that it views everything else on this planet—as yet another manifestation of a world separated from its source, itself, and each other. We use the PLO only by way of example and because its members are using *our* acronym.

The real PLO's purpose is to assist in truly liberating the planet. This PLO is here to see that the last chapter of history is written. Rome is to fall for the final time, and history is to fall with it. That is what is meant by the prediction that the world is destined to come to an end.

The PLO—Its Historic Position

Your world views history as a record of everything important that has ev er happened. Nothing could be further from the truth. History is little more than the distorted chronicling of endless human ego posturing whose sole purpose has been to reinforce the state of denial. Since the planet has decided that "no" is no longer an acceptable answer, historic times have no choice but to come to an end.

Unless you are an awakened member of the PLO, this concept may be difficult to swallow. For the benefit of those who are not yet fully awake, Mission Control offers some examples of history's inherent flaws that are leading to its timely termination. These examples will probably be equally difficult to swallow.

Exhibit A: You have no doubt noticed that this planet's civilizations rise and fall with remarkable regularity. Historical accounts of these events invariably explain the reason for the decline. In the case of the demise of Rome, historians point to such contributing factors as moral decadence and a rather unfortunate epidemic of lead poisoning. However, these were merely the symptoms, not the cause, of Rome's ruin.

The true reason for the nauseating up-and-down motion of all human civilizations, Rome included, is that their ideologies, political systems, and social structures have failed to liberate anybody—especially themselves—from the vicelike grip of fear. Civilizations fall for one reason only: They are all built on fear and denial. Their subjects subsequently lose themselves in orgies or wine and dine themselves on lead to either distract themselves from that horrible fact or to get themselves out of it as fast as they can. And, because the real issue is never faced, human bondage continues uninterrupted from one civilization to the next, ensuring the ultimate collapse of each one in turn.

History isn't exactly repeating itself—it is stuttering over an issue it hasn't addressed.

Americans are another excellent example of this repeated dysfunctional descent into slavery and collapse. Duped by their Declaration of Independence, they fancy themselves to be free. But having a voice in where the next nuclear plant will be built instead of whether it should be built, being able to eat anything they wish from a totally infected food chain, and having the inalienable right to file for an extension on paying the taxes that are being used to kill them are not exactly the freedoms that the authors of their constitution had in mind.

The fact that their personal prisons may be expensive, tastefully decorated, and equipped with a stereo, a TV, and a VCR does not make their confines any less a cell. Trading life for economic survival is not liberty; liberty is freedom from both fear *and* survival. That was the freedom that was originally intended to flourish in the United States. In truth, the author behind the authors of the American Constitution was Spirit—not the Bank of America, the Federal Reserve, or the IRS. America, by acting out of fear, is busily turning its back on its own destiny and facing an impending crash. However, it

54

E.T. 101

is in good company, because the rest of the world is doing the same.

The Planetary Liberation Organization is a manifestation of Spirit, and it is here to see that America and all the countries of this world turn and face their spiritual future instead of culturally caving in under the weight of denial. The PLO is here to liberate the Earth from its dysfunctional historic repetition of destruction and decline.

Exhibit B: The historic period has done little but propagate lies. Even if an historic account is accurate (which it usually isn't), the event that it chooses to describe is nothing more than the acting out of a fundamental lie. Consequently, whatever truth the situation may have held is invariably missed.

An example of this can be seen in American history's rendition of its bloody conquest of the West. Every account will tell you that the white man won his war against the heathen and savage Indians. Not only are these reports unabashedly biased (and the lie of separation that fueled the event carefully hidden), but the fact that the allegedly defeated Indian Nation was the true victor is never acknowledged at all.

You read that sentence correctly. The Indians won their war with the white man. That war was a struggle for spiritual ascendancy, not a battle to determine who would subjugate the land. The Indian peoples, who are fifth-dimensional representatives, sacrificed themselves with their very blood to assure that this nation would become the spiritual giant it was meant to be. History books are incapable of recording that fact because their function is to meticulously gather material in support of the fear-filled national egos they represent. Because they serve human self-deception, historic accounts totally ignore the grander and deeper movement of Spirit that is always the only truth behind any fact.

The Planetary Liberation Organization is here to assure that the fifth-dimensional victory the Indian Nation actually won is finally demonstrated, and that the third-dimensional wounds sustained in that battle are fully healed. America will then assume its true spiritual identity and discover what "Manifest Destiny" really means.

⚡

Note: For those of you who are having trouble digesting the truth of this statement, let us point out another historic example. If you believe that Japan lost World War II, we would like you to take a look in your driveways.

The PLO and Armistice Day

The end of history should not be perceived as a frightening event. History has been the frightening event. Its end is the liberation that will exalt humanity, not strip it of its power. As the planetary ego aligns with Spirit, it will have no further need to defend itself against an enemy that does not exist. Nor will it have any use for carefully recording that process and then passing it on to future generations as an impaired model for them to imitate.

The Planetary Liberation Organization is here to cut an energetic pathway to a state of grace. In doing so, it will enable the planetary inhabitants to claim their freedom from both their national and their personal histories, as well as allow them to experience their true spiritual and physical wealth. The celebrations that marked the endings of your world wars will look like sedate little tea parties compared to the global celebration of true peace and liberation that will date the end of historic times.

The United Stellar Army Corps of Engineers

Many of you are specialists in fifth-dimensional technologies and are members of the United Stellar Army Corps of Engineers. This organization has nothing to do with the other U.S. Army Corps of Engineers, as their technologies are strictly 3-D and will be useless shortly. The United Stellar Corps members are masters of transmutation, high-frequency vibrational medicine, and dysfunctional disarmament. The Corps is here to bring these and other technologies onto this plane to facilitate the physical and spiritual alteration that is currently in process.

The 3-D Dilemma: Third-dimensional sciences are ill equipped to handle the awesome shift that is now taking place because they are not true sciences. They are little more than elaborate systems of measurement that have been used to dominate the environment through a faulty understanding of energy.

The major reason human sciences have been unable to evolve beyond the level of pseudosciences is because they are based on fear and manipulation rather than on love and creation. Another reason they currently fail to discover anything worthwhile is because they have been bought off by the industries that control their funding. These industries have a vested interest in not developing anything that would liberate this planet from economic slavery. Therefore, they have created a parity program that is actually paying scientists not to discover anything that might overthrow the current system. As a result of these factors, the sciences have been cornered into an unempowered position that has successfully placed a cap on any further third-dimensional scientific achievement. This has largely eliminated their having any meaningful input into the transmutative process.

The 5-D Response: Because of the tacky situation Earth scientists have gotten themselves into, the majority of the United Stellar Corps is not among their ranks. Cleverly disguised as housewives, office workers, hairdressers, cab drivers, or mild-mannered reporters, they stand outside the scientific domain, ready and waiting to be called into action. That call is now being sounded.

Unlike its third-dimensional counterpart, the United Stellar Corps is staffed by true scientists. The technologies of these scientists are based on love, manifest light, and are true acts of creation. Because their sciences are creative arts, they have no need to systematically disassemble life, destroying it in an attempt to see what makes it tick. They instead participate with life to create new options where those options didn't seem to exist. A major function of this corps is to offer free passage on the 5-D express train they engineer, enabling the planetary inhabitants to disembark from the 3-D local that has them rattling slowly around their cosmos riddled with fear.

The technical representatives of the United Stellar Corps are here to usher in the final frontier. That frontier is creation, and entering the frontier of space is but its natural consequence. The key to interstellar travel lies in mastering the creative process, not in acts of technological conquest. That mastery will lead the planetary inhabitants out of their stupor. And, as a result, it will also lead them out of their solar system, their galaxy, and their dimension.

The United Stellar Army Corps of Engineers is here to open the doors to technologies as yet undreamed of. These technologies are the legitimate offspring of Spirit, not the disinherited children created through the unholy marriage of manipulation with the frightened machinations of the mind. The Corps is here to unlock the gateway to the stars, and to escort the Earth and its inhabitants to their rightful and royal place among the vast creative forces of the universe at large.

Adult Children of Dysfunctional Earthlings (ACODE)

Adult Children of Dysfunctional Earthlings may not sound like a job category, but it is—and you are all in it. Some members of the mission, however, are especially gifted in the art of reawakening and are experts in assisting other recovering Earthlings to do the same. They are called ACODE First Class, and they are experts in the field of owning up to their true magnificence.

This advance guard of ACODE is noted for demonstrating dramatic changes in consciousness at very rapid rates. You will know you are among them if the picture on your driver's license begins to look like it was taken during a past-life reading or like a photo someone snapped at a masquerade party you don't recall attending.

Members of ACODE are also extremely adept at the "cosmic-quick-weight-loss diet." Although their physical bodies may or may not reflect the effectiveness of this regimen, their auras always do. The first and only step in this program is to stop hiding. ACODE members have the ability to do so overnight—and wake up laughing about their social security numbers, their mortgages, and especially their alleged careers. They have the singular ability to burst into their full presence and assume their sovereignty without even giving their friends and colleagues two weeks' notice. As a result of this skill, they are indispensable in blowing apart the dysfunctional games of everyone else around them.

Adult Children of Dysfunctional Earthlings is the front line of Mission Control's Planet Renewal Project, and their support groups are the heavy artillery of human liberation.

The MASH Units

The Mobile Astral Surgical Hospitals, or MASH units, are stationed throughout the planet as well as off-planet. Their medical personnel are masters of interdimensional internal medicine.

Like the United Stellar Corps of Engineers, the members of these units are, for the most part, not to be found among the established medical profession. The biological knowledge that the process of transmutation requires is beyond that profession's scope and, luckily, beyond its treatment programs. If, by some fluke, medical doctors are able to diagnose transmutation correctly, they will probably try to develop a vaccine to stop it or a drug to squelch it. Mission Control does not recommend you take either. If symptoms persist, we advise you to consult your local interdimensional physician instead. To do so, go within yourself and ask for the MASH unit emergency line. Once you have been connected, request the assistance you need. A MASH unit member will respond to your call on one dimension or another.

Transmutative Symptoms: One of the most common transmutative symptoms is exhaustion, usually resulting in a phenomenon called "vegging out." This is only natural because the physical body is rearranging its cellular structure to accommodate its culinary shift from burgers and fries to a diet of pure light. Do not be alarmed by the resulting fatigue of this process. It would be more alarming, given the condition of your food chain, if this change did not take place.

Other commonly reported symptoms include the discomforts of "transmutation fever" that the emotional body reports as it sees itself being dismantled, fumigated, and remodeled. Don't give this condition undue attention, since the emotional body is largely opposed to all of this transmutational stuff and tends to complain constantly. For similar reasons, the mental body may inform you that it is suffering from terminal confusion as it watches Spirit assuming command and imagines itself about to be fired. This is, of course, untrue, and it may be useful to remember that the mental body loves to misinterpret and misrepresent.

If that ploy doesn't work, the mind may also point out that suddenly it can't remember anything, as though this were proof positive that transmutation is terribly dangerous and grounds for your immediate retreat. Disregard the report. The truth is that the past is being removed as a mental reference point and replaced by future ecstatic models. The mind can't remember anything because all useless data is being culled from its files. What it perceives as imminent danger is actually its imminent liberation.

Lastly, the ego and the personality will undoubtedly come up with a litany of complaints that are hair-raising if you give them any credence. Do not listen. They, like the mind, haven't a clue about what's going on and are inaccurately reporting their reconstruction through their fearful sense of being abandoned.

Although most of the symptoms you may experience will not have very much biological substance, they can still be uncomfortable and unnerving. Your old life is dying and your new life is emerging. That process can cause quite a bit of physical, mental, and emotional disturbance. If undue discomfort occurs during the process of your birthing, check in to one of the recovery rooms staffed by our MASH unit members. They specialize in assisting you to safely fall apart on all levels.

Transmutative Cures: There are no real cures for transmutation, nor should you desire that there be, because transmutation *is* the cure. It is a natural process that is absolutely necessary for the next step in your evolution. It is also the only way you will be able to withstand the increased vibrational field of energy you are about to enter. The closest thing to a remedy for this process is to willingly allow and assist these vital changes.

Voluntary consent is the jurisdiction of the "patient." Assistance is in the competent hands of the MASH unit members. Their personnel have one primary function: They are here as highly trained professionals to assist in the planetary birthing process. They are interdimensional midwives sent in to help the planetary population and the planetary sphere through the trauma and pain that may accompany this awesome act of spiritual awakening. They are specialists in the transmutative procedure and are equipped to handle all the psychic, emotional, and physical complications the emergence of new life can entail. All MASH units, both on- and off-planet, are ready and awaiting all calls that will bring them to the delivery rooms immediately.

(For details on some specific surgical procedures, see "Interdimensional Brain Surgery" and "Emotional Body Exploratory Surgery" in *Assistance*. If you are stationed in Canada or the United States, you can contact the MASH units by going within and dialing 911.)

The Code Talkers

In much the same manner that the Navajos used their native language to outwit the Japanese during World War II, the Code Talkers of this mission have a similar task and technique. The major difference is that the directive of the Code Talkers has been altered since 1945. Now their responsibility is to get information *to* the planetary inhabitants instead of trying to get it *by* them.

Using their indigenous tongues, the more recently arrived Code Talkers are able to transmit vital interdimensional information to the cultures that surround them while bypassing those cultures' linear minds and models. And, just as Navajo was Greek to the Japanese, the concealed communications of the Code Talkers are equally inscrutable to the locals and will likewise never be deciphered. Mission Control can say this with absolute certainty, because cracking a code implies being able to translate its information into English or some other linear language. In the case of this code system, that would be like trying to force a three-dimensional object onto a two-dimensional plane. It can't be done. The fact is, Mission Control could even go as far as to blatantly broadcast the key to the code, and still no one would get it, because the tool they would be using to get it with would preclude their getting it at all.

Code Talkers are cleverly positioned throughout the planet, and their function is critical to the success of this mission. For that reason, Mission Control will not release any further information concerning their whereabouts or specific tasks. This is for their protection, as we cannot be assured that some "nut" isn't reading this manual who may consider code talking to be an un-American activity that rightly deserves another national inquisition. Likewise, we cannot be certain that some old-age entrepreneur may not also be reading this material and be struck by this golden opportunity to charge everyone for this communication. Although any attempt to interfere with the Code Talkers would ultimately be futile, Mission Control is being conservative in this matter because we dislike any interruptions on our lines.

If this necessary absence of details and specifics leaves you unsure whether you are a Code Talker or not, a clue may lie in your response to this manual. If it makes total sense to you and you intend to file it on your bookshelf somewhere between *The Wonderful World of Macramé* and *Chilton's Complete Repair Guide* for your car, it is very likely that you are a Code Talker here on special assignment.

The Interuniversal Banking Community

The members of the Interuniversal Banking Community are here to assist in dismantling the dysfunctionality of this world's economic structure. They are artists in the use of plastic currency and are armed with true MasterCards, unlimited credit, and the knowledge of how to charge everything.

The basis of their mastery lies in their complete understanding of the plastic nature of reality itself. They are not confused by form; they create it. They know that their attention and fascination are the foundation of the reality they build; therefore, they specialize in withdrawing their consciousness from dysfunctionality and placing it on spiritual truth instead. As a result of their efforts, we will shortly be in a position to present the planetary inhabitants with a viable economic recovery program.

This program entails realigning this planet with the system of divine economics that the majority of the universe currently enjoys. The activity of the Interuniversal Banking Community will assist in the planetary realization that no life needs to earn the right to live. This is not a statement of economic heresy; it is a statement of liberation based on truth. Once the human population has received that medicine, survival will crumble, and unending abundance will flourish in its place.

Carte Blanche is now being extended to the entire human species by the Interuniversal Bank, and its bankers are here to approve unlimited human credit. Mission Control suggests you apply soon.

The Intergalactic Board of Realtors

The members of the Intergalactic Board of Realtors specialize in the reacquisition of all planetary realty for future development as spacefront property. As the Board takes over, many of the current landlords will be evicted. This takeover does not imply that the Third World is about to invade your countries and occupy your shopping malls. It means that survival is over. The old world's systems are in collapse. Those who wish to continue in those systems will be graciously asked to leave, because their motivator, fear, is being relocated to another planet where its subdivisions are still welcome.

The Intergalactic Board of Realtors has already placed this planet on an Interdimensional Multiple Listing for recolonization by the Forces of Light. The Board is present on this plane to see that this planet's personal property frenzy comes to its natural end and that the illusion of planetary ownership is replaced by the legacy of planetary stewardship. That is the birthright that was intended. That is the heritage that will result in true equity and abundance on this precious planet.

Light is the rightful inheritor of the Earth. Our realtors are here to ensure that the terms of the contract are drawn up properly as the deed is transferred to its legitimate heir.

The Cosmic Computer Jocks

Some mission members are in our special computer division. These members are experts in the art of interdimensional interfacing. They are this mission's Cosmic Computer Jocks, and they have the capacity to act as the very linkage between the dimensions.

This division specializes in the translation of third-dimensional binary computer language into a fifth-dimensional unary linguistic system. Although any computer expert on this planet will tell you that such a translation is not possible, what they actually mean to say is that they could not do it. That is why they are not being asked to. We have sent in our own specialists instead.

Another aspect of this division's task is to bring new software to this plane with programs that no one here has dreamed possible. These programs are fifth-dimensionally designed and apply to every aspect of the transmutational process. Not only can they unscramble existing confusing and dysfunctional programs, but they can also realign them with our database which will automatically reprogram them back into light.

Our computer team is here to disseminate our new software throughout the planetary sphere in preparation for the final dimensional link-up. Their very bodies are the silicon chips of our computer matrix, and their presence is the keyboard of our system. They are state-of-the-art hardware and are completely immune to any computer virus as well as very user-friendly.

The Rainbow Warrior

The Rainbow Warrior is a warrior of the Spirit, and every member of this mission is a Rainbow Warrior. In fulfillment of Native American prophecies, inter-galactic and interdimensional forces have gathered on this planet at this time to liberate her in the name of Spirit.

The Indian peoples are fifth-dimensional emis-saries, equipped with their own private lines to spiri-tual truth. As many tribes have predicted, a powerful spirit is now returning to the Earth to lift her out of her decay and despair into a new and glo-rious realm ruled by the Spirit of Love. In the Na-tive American vision, the term "rainbow" expresses that this is to be a global event not limited to tribe, nation, or race. The greater truth is that it is not even limited to this planet. All Universes of Light have sent in their representatives to help cut the passageway to this incoming era of liberation and life. The rainbow that they represent is far beyond the spectrum of light that the people of this planet have ever seen, and its colors are much richer and more vibrant than the Earth's shadowy light has thus far been able to reveal.

The Rainbow Warriors are a living expression of the new light, sent forth by Mission Control in honor of the Native Americans, their prophecies, and all their relations. The great Universes of Light congratulate and salute our Indian delegation on a job well done. The Kingdom of Light they foresaw is dawning even as you read these words.

The Royal Celestial Air Force

The Royal Celestial Air Force is many, many times vaster than all the national air forces of this planet combined. It is a division of the High Command of Light. As such, its strength is greater than any military force this planet has ever beheld.

Every major head of state on Earth has been advised of our presence and assured that none of their little Star Wars devices will ever have an opportunity to be fired. The reason you have not been informed of this fact is because your military forces are beside themselves with the loathsome idea that their pop guns are meaningless. They also do not wish you to know exactly how useless their environmentally crippling military expenditures actually are. To put it bluntly and in Earth terms, they are protecting their own asses.

Our presence in your planetary arena is not for invasionary purposes. If we had wished to invade, we would have done so long before now. We have girded this planet with our ships to form a resonant field that assists the Earth in the transmutative process. Our craft are also here to protect and communicate with our ground crew members who are facilitating the transmutative process on the planet's surface. Our crew are continually monitoring our ground personnel for data. This is to aid them in their awakening process and to assure that they arrive at their respective positions on time. We have the ability to recall mission members to our craft for instruction or assistance, and we do so constantly.

Although the power of the Royal Celestial Air Force is greater than anything the inhabitants of this planet have ever seen, our love is also greater than anything these inhabitants have ever dared to dream. The Earth is not endangered by our presence; she is exalted, for we are here to assist in the breaking of her bondage and to fulfill her regal destiny. Our squadrons stand at her side and at her service, assuring her safe delivery into the light.

We are the Royal Celestial Air Force, in service to all humanity out of our love.

✝

(This article is a translation of a direct transmission from the Commander-in-Chief of the Royal Celestial Air Force.)

The Quark Alliance

As a member of this mission and its planetary transition team, you should be aware of the presence of the Quark Alliance. Although you cannot join this alliance, it is joining you, so we are including a short explanation of its function in this section of the manual for your information.

The Quark Alliance is a very powerful organization whose work is not immediately obvious because its jurisdiction is subatomic. Its presence and activities are dismissed by the human scientific community because any admission of its existence would force them to recognize intelligence in places that would frighten them to find it. Acknowledging this Alliance's presence would also debunk science's high priesthood by challenging both its knowledge and its control. Since the scientific community is unlikely to defrock itself voluntarily, its denial of the Quark Alliance is apt to continue.

The Quark Alliance has been responsible for many recent technological failures which have been falsely attributed to such things as human error or, on occasion, metal fatigue. Although human error should never be underestimated, the issue of metal fatigue only hints at what is actually happening. What is occurring is conscious communication at the atomic level that has resulted in a unanimous decision to alter the fabric of your physical world.

Metal is not becoming fatigued; molecules are. In fact, they are not just fatigued—they are entirely fed up. They are no longer willing to be servants to a technology wielded in denial of life and to the planet's jeopardy. Consequently, they are flat-out refusing to cooperate, creating a certain amount of technological havoc. Through the work of the Quark Alliance, atomic particles have begun their realignment with the Forces of Light and are in the process of rearranging physical reality as they pull the subatomic carpet out from under the feet of denial.

We apprise you of the Quark Alliance's existence so that you will not be surprised when you see the fabric of modern physics unraveling before your eyes. Pay no attention to the barrage of technobabble that you will undoubtedly hear as science attempts to maintain its power in the face of its ruin. In fact, you may as well just sit back and enjoy it, knowing it is only the passing protest of old form cracking under the superior force of the incoming light. What looks messy on the surface now will soon give way to a new order filled with harmony, cooperation, and joy.

⚡

As we have mentioned, this list of job descriptions in no way represents the full spectrum of the mission. These are only brief descriptions of some of the tasks that some of you chose to do. The actual depth and breadth of the mission is beyond human description because it was architected in another dimension. On this dimension, you will have to be satisfied simply knowing that Mission Control is never sloppy. Our plans and programs cover every life form on this planet. So, proceed with your specific mission, secure in your purpose, and strong in your love—and don't forget to keep in touch.

4

Troubleshooting

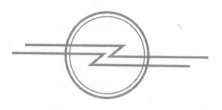

MISSION CONTROL acknowledges that the process of waking up is a little tricky. Even though you are genetically encoded to do so, by the time you reach the point of activation, you will be totally convinced that you are an Earthling. You will most likely be exhibiting their worst characteristics plus wearing any number of their scary disguises. You may find yourself in the middle class, a self-made man, a self-denying woman, terminally confused, completely content, following a guru, joining gun clubs, sweating your mortgage, watching TV, defending your nationality, owned by your corporation, taking care of your lawn, dialing for dollars, a "victim" of religion, seeing a shrink, jogging in circles, doing lunch, an attorney and/or working for the DOE.

This is, of course, a very partial list of the frightening possibilities. It can be summarized by saying that you will have been successfully brought to your knees—not out of reverence for life, but out of the unending effort of scrambling in slavery for survival.

Fear, in any of its many forms, will probably have managed to topple you in one way or another.

In addition, many of you will be in your forties and over the hill. (Remember, the majority of this incarnational group entered shortly after World War II—see definition for "Crawl-Ins.") This means that you may have had many medals, awards, bowling trophies, and degrees bestowed upon you (depending on the level of slavery you bought into), plus all the power, position, and credit cards that were held out like carrots to further buy you off. And, in the worst of all possible scenarios, you may also have a white male body that lives in Wilton, Connecticut, and has mistaken its portfolio for its identity.

To all this, Mission Control says, "Yikes!" We also say, "Thank God you're Christ."

What follows in this section is a little helpful advice in areas that are commonly problematic and typical of dysfunctional planets. Although simply and totally waking up would eliminate most trouble spots, we are aware that many of you cannot do so overnight because of the degree of brainwashing you have sustained. Our advice, however, is please don't drag this process on too long or you may miss the mission entirely. This is Mission Control. Carry on.

Passing for White

As you begin to awaken interdimensionally, Mission Control advises discretion. Keep a low profile and act as "white" as possible, unless you happen to be Chinese. (A little common sense is useful here.) This is for your own safety. Don't forget, the cultures of this planet are built on fear. They fear everybody, everything, and *all* differences; moreover, they kill in defense of those fears.

Up until now it has been largely unnecessary for Mission Control to caution you on this matter, since you have not had a clue as to who you are and why you are here. However, as you begin to sense your true identity, be extremely careful. For instance, going up to someone and casually saying, "Hi, I'm from Sirius—I understand you're a native," will *not* win you any friends or influence many people. If you're lucky, they will just think you're nuts. If you're unlucky, they may commit you.

Remember, you came here to dismantle fear, not to elicit it, so be cautious about cocktail conversation and try not to alienate the aliens.

Closet Cases

Although you are in some danger from the indigenous population, the greatest danger you face is from other extraterrestrials who refuse to awaken. The local alien population, for the most part, will be inclined to dismiss the claim that extraterrestrials are in their midst by the millions as a crock of biodegradable matter. They are so certain they know what is happening that they will probably miss what is happening until it has already happened. Because of the one-dimensional nature of their belief systems, the natives will be unlikely to launch any witch hunts.

On the other hand, extraterrestrials who are bucking their genetic coding are a bit more dangerous and should be approached with caution. They are more likely to strike than their cocksure counterparts. And, if any witch trial appears on the docket, they will undoubtedly have placed it there, as well as appointed themselves judges.

The Messiah Complex

As you are awakening, there are some pitfalls we would like you to avoid. The most important one is the dreaded Messiah Complex. Before mission members are completely on their multidimensional feet, this complex tends to have some appeal. Mission Control would like to take a moment to make it a little less appealing.

Being Christ and thinking you are Christ are two different matters. If you only think you are Christ, you will then act like you think Christ would act, which usually entails trying to save someone.

Let us make one thing very clear: This mission is not about "saving" anyone. All inhabitants of this planet are masters. Even the aliens are masters who are here doing a brilliant job of mastering being aliens. Everyone on the planet knows the game, and everyone has made their decision. If a person has chosen to continue as a master of limitation, that is his or her inalienable right. Saving people from their rights is not the intention of this planetary mission. And having our ground crew members running around with messianic fervor trying to rescue people from their free will is not through any request of Mission Control nor by any mandate of the Councils.

The Earth has elected to evolve beyond limitation; however, anyone who opts to explore that process further is free to do so—just not on this planet. Such people will be allowed to continue their experiments with limitation on some other piece of planetary property that is at a less advanced stage in its evolution.

The members of this mission have chosen to master divine expression instead of limitation, and are being asked to do so on this planet at this time. It is critical that you remember that one choice is not better than the other; one choice is just more suited to this planet than the other. Do not, in your half-awakened state and out of misdirected zeal, attempt to "convert" anyone to the choice you have made. Instead, *be* the choice you have made.

Mission Control expects our members to respect everyone's sovereignty and decisions. We also expect you to stand in your full presence and emanate your divine essence. In this manner and in no other, you will have the power to effect another's choice to do the same. Your embodiment of Spirit is the only act that will assist the mission in unfolding smoothly and efficiently to its destined conclusion.

The Burden of Spiritual Significance

The Burden of Spiritual Significance, like the Messiah Complex, is a trap we advise you not to get caught in. The problem with Spiritual Significance is that it is a by-product of spiritual ambition and, as such, it would best be your spiritual ambition to avoid.

Acts of spiritual ambition are, by their nature, devoid of Spirit. They will only result in separating you from Spirit and, therefore, the mission. This is not to say that we don't expect you to do anything of any spiritual consequence while you are visiting this planet. We *do* expect you to have a spiritual impact here, otherwise we would not have sent you in. However, becoming entangled in the "importance" of your acts will lead you into an identity that is less than who you are.

You are here with one primary directive: to embody the Spirit you serve. If you allow yourself to become sidetracked by your "spiritual significance" and lose yourself in the "grandeur" of who you are, you will simultaneously lose track of your *real* significance and fall short of this mission's goal. Remember that you are here to become a living expression of Spirit. Nothing you will do or say is an acceptable substitute for becoming who you truly are.

The Chicken/Head Syndrome

As dysfunctional patterns are being dismantled and fear is being unceremoniously kicked out of the driver's seat by Spirit, you may experience the chicken/head syndrome. (Our sources indicate that you have chickens on this planet—indigenous birds who are noted for running around after their heads have been cut off. This is our first exposure to chickens, but we find their behavior useful, so we have renamed this syndrome in their honor.)

The chicken/head syndrome refers to the neurological phenomenon that a beheaded chicken experiences when its body continues racing around frantically as if something were still in control. This goes on for a short while until the neurological circuitry catches up with the fact that the bird is officially dead. This is precisely what can happen when fear is eliminated from your systems. Fear's neurologically patterned behavior may continue marching around for awhile, acting as if fear were still in charge.

You have two options in dealing with this condition. You can treat this vestigial behavior in the same way we have noticed you treat flies. (This is also our first encounter with flies, but they seem to be just as useful as chickens.) You may allow them to buzz around until they drop of their own accord, or you may swat them and get it over with. The only thing you should never do is identify with them.

Fear and its patterned behavior is not and never was your identity. Fear is a parasitic life form that no longer has any biological business being on this planet. If it is helpful, think of fear as a fungus from outer space that successfully invaded eons ago and has been hosting off your systems ever since. Fear no more defines your being than a case of athlete's foot defines your body. So, whatever course of action you choose to handle this syndrome with, remember that it's almost over and you're not it.

Integrity—
Its Care and Maintenance

As a "crawl-in" to this mission, you, by definition, have some pretty big handicaps. As you are asked to stumble out of your wheelchair and into an upright position, you may encounter some enticement to remain seated and rest on your handicapped privileges. Mission Control would like to take this moment to assist you to your feet.

The biggest handicap that you suffer stems from the fact that this mission demands total integrity, while the cultures you represent demand little or none. The reason for this is that Earth cultures have one basic thing in common: They are all dysfunctional. Once a culture has decided which dysfunctional aspects it wishes to represent, it raises a flag to declare its position, packages its preferred brand of dysfunctionality for consumption at home and abroad, and passes it off as a national heritage to be proud of and protected at all costs. Because you have to claim some nationality in order to get in here, none of you has been spared an identity that is at least a million light years and exactly 180 degrees off from the truth.

The temptation to remain dysfunctional arises from the fact that it has been such a thorough and arduous journey getting there; somehow, it feels wasteful to just chuck it. Because of this illusion of waste, you may find yourself clinging to false identities or co-dependent relationships that prolong the recovery act. These double-dealing relationships, whether with yourself or others, are based on a dysfunctional complicity that thrives on an unstated request. That request can best be expressed as, "Please don't disturb my sense of limitation. It may be Auschwitz, but it's home."

The problem with maintaining this "pact" is that you cannot pass through the doors of the fifth dimension lugging dysfunctional baggage, and there is no handicapped parking nor any wheelchair access. All false identity must be relinquished at customs where your belongings will be rifled for contraband states of consciousness. These contraband states include dishonesty, manipulation, any and all feigned limping, refusing to relinquish your survival identity, and every hidden, unholy agreement that was made out of fear and denial.

Mission Control is aware of the courage that re-alignment with the truth requires, but we are also aware that no one will be successful in any attempt to smuggle a lack of integrity across the frontier of the new incoming civilization. Be gentle with yourself and with others during the time of your rehabilitation, but also be scrupulous in this matter because there is no room for deception. And remember, giving up your crutches willingly now is far preferable to being busted at the border.

Discernment

At this time of transition, be very careful about who and what you are following. In fact, if you are following at all, that is the first indication that you are off track. For those of you who are still the students of gurus, we recommend discernment.

This is no longer the time of great spiritual teachers. It is now the time of great spirits instead. This shift from master/student to just plain master may cause a temporary unemployment problem in India and elsewhere, but do not be alarmed. The true masters of light will make the shift with ease and will welcome your upcoming graduation with the same relief that they welcome their much-deserved retirement.

Others of you consider gurus passé and are following disembodied channeled entities instead. Again, we advise discernment. Many of our forces have gained entry to this plane through the use of channeling. However, we are not the only ones who have gained entry this way. There are many disembodied energies who are masquerading as the light and throwing their confusing two cents into the global pot. Being without a body is not an instant membership card granting the bearer status among the Forces of Light. There are an inordinate number of entities running loose right now, channeled and otherwise, who have no bodies and are solely interested in an opportunity to use and abuse yours. An important key in dealing with these energies is to approach as a master and not as a student. If you stand in the truth of that identity, you are very unlikely to fall for a lie.

Always test the energies you are in contact with to make sure they are not just fourth-dimensional freeloaders with a predisposition to remain in the dark. If an entity shuns the light and avoids standing in its presence, acting somewhat like a vampire who has just been confronted with a crucifix, you can be fairly certain they are not in the service of the Forces of Light. Anything that cannot tolerate the light is not assisting the light and should be taken *to* the light as soon as possible. Check your behavior and thought-forms as well. Much of what you have considered to be the product of your personality and upbringing may actually be the behavior of a fourth-dimensional entity that is time-sharing your body.

This is a particularly important issue right now because there is a great deal of disturbance on the fourth dimension which is leaking into the third. (Please see the second note under "UFO" in chapter 1 for more information on the dimensions.) This is the unfortunate result of a little interdimensional misunderstanding. As the fifth dimension continues on its spiritual descent into the third dimension, it is now passing through the frontiers of the fourth. Some of the fourth dimension's darker denizens believe this incoming light to be a threat, and mistake their imminent transformation as a serious assault. A number of them have consequently formed a resistance movement that is fighting back, even though we are not fighting at all. The temporary chaos this has caused is making the fourth dimension look like a bad brawl in the *Star Wars* bar, and some of its disembodied refugees have made their way into the third dimension. Learn to recognize these energies and stay clear of them.

If a disembodied entity manipulates you in any way or wants your following at the price of your freedom, that entity is not on our team and doesn't have your best interests at heart. Any energy that is not contributing to the realization of your magnificence and mastery is not a part of this mission and is in the service of the dark forces. If an entity fails to meet this criteria, escort that being into the light.

Through your alignment with the light, you are in a superior position in relation to these temporarily confused forces. You have the power to bust them and lovingly usher them into the light. You can do this by identifying the entities, breaking any agreements you may have made with them, and, by an internal visualization, leading them into the light. This act will serve to assist the mission in its peaceful and efficient descent through the fourth dimension, while hastening its awaited arrival on the third.

Do not misinterpret this information. It is the *light* that is superior to these fourth-dimensional forces and not your winsome personality. Pitting yourself against them as if *you* were superior will invariably end in your resounding defeat. Call upon the Force of Light in all your dealings with these energies. Your success will then be guaranteed, and there will subsequently be less mess to clean up.

⚡

Note: Some of our Special Forces units have published materials containing technologies to assist you during this crucial transitional period. If you are interested, please write to us at Mission Control and we will see that you receive information about their publications. Our address appears on the "Extraterrestrial Census Form" at the back of this book.

Landing Instructions

Some of you are in such a state of shock from finding yourselves in the third dimension that, in protest, you have refused to land. Mission Control would like to point out that you are useless to the mission if you are still circling the planet in a holding pattern. We would also like to point out that it was *your* choice to sign up for this mission, not ours.

From your frantic transmissions, we have gathered that you are nervous about catching whatever it is "they" seem to have on this planet. Although we understand your anxiety, we would prefer to discuss your imminent danger *after* you have made your landing.

Technically, Mission Control cannot interfere with your free will; however, we can reassign you. You may be transferred, if you so wish, to another dysfunctional planet. Unfortunately, most of the positions we have open right now actually make this place look good. The mere mention of the possibility of re-upping for boot camp on Planet X is usually enough to coax most of you out of the skies and on with the mission. However, if you are still unwilling to make your approach, please contact Flight Control. Maybe they can talk you through a landing.

Culture Shock

Culture shock is unavoidable as you begin to awaken to our presence as well as your own. Although you already are veteran travelers of the dimensions, your true identity will be a news flash to your third-dimensional consciousness. The impact of recognizing your multidimensional nature will send ripples of apprehension through your limited sense of self, giving the prospect of a sudden, underfinanced move to Calcutta far greater appeal. Even though it is only the security of your insecurity that is at risk, try telling that one to your emotional body. The emotional body may be more inclined to fling itself off a cliff than deal like an adult with this incoming vibrational shift.

Culture shock is temporary, but we mention it so that you can prepare. And while you are at it, get ready for the additional shock that Mission Control is populated by a largely nonhuman staff. The human race is a root race that extends throughout the worlds, but it is only one of many. For a people who have not yet adjusted to the differences among their own kind, our presence may seem an alarming act of brotherhood you are being asked to face.

If it is any consolation, many of you on this mission are merely *disguised* as humans for the sake of this planetary transition. We hope that information helps your personality make its adjustment, even if it doesn't exactly cheer up your beleaguered emotional system. Also, keep in mind that the culture shock of awakening multidimensionally is nowhere near as dreadful as the shock you felt when you first woke up to find yourself here.

The Yo-yo Effect

The yo-yo effect is a name we came up with to describe the bodily and emotional changes you may feel as the transmutative process kicks into full gear. Cellular transmutation is necessary to accommodate your evolutionary leap into light, but since this process is physical, it has some attendant symptoms that you might as well get acquainted with.

There will probably be moments of exaltation as you feel the rush of incoming light entering your systems. However, these are often followed by sudden crashes of energy—crashes that can be felt by the body, the emotions, and the mind as you temporarily swing back into the old reality. Do not mistake this for manic depression. It is only a simple case of ecstasy followed by your denial's insistence on returning to the pits it knows and loves so well.

The accompanying physical symptoms will differ from person to person, but any combination of aches and pains is possible including the discomforts of nervous disorders, and fatigue may set in. Mission Control does not suggest you rush around to every doctor in town trying to figure out what you've got, unless you have unlimited funds. Our advice is simple and about as good as you'll get. Be kind to yourself. This is an enormous shift. If you have trouble, just take a couple of light pills, go to bed, and call us in the morning after you wake up.

⚡

For more information on transmutational symptoms and cures, please refer to "The MASH Units" under *The Mission.*)

Deployment of Troops

This mission has an overview and an objective; however, it does not have a battle plan. One reason for this is that we are not in a battle. Another reason is that all our movements are directed by Spirit and change constantly in response to Spirit's requests. For this reason, you must also be willing to alter your plans in accordance with Spirit and go where you are summoned at a moment's notice. What was true yesterday may not be true tomorrow. You must learn to rely on Spirit for all your up-to-the-minute instructions.

This reliance on Spirit is the mission. It is also your direct line to Mission Control and all its forces. (Remember, Mission Control is *not* an outside authority. We are a service, both internal and external, that strongly recommends you do not look to outside authority for your counsel.) No one but your divine Spirit can tell you your truth, where you should be, or what you should be doing. Spiritual self-reliance is the essential shift in consciousness that the mission is here to help implement. It is also the shift you personally agreed to make on behalf of this planet. We exhort you to be ready. Be awake and listen, because the troops of Spirit are now being deployed.

Emergency Procedure

In the event of an emergency, Mission Control has reserve forces ready to swing into action. We also have the ability to simultaneously activate all the genetic codes of our ground crew members, and to instantly call you to your respective positions. As we have mentioned before, this is not a preferred course of action because of the shock of the procedure. Many nervous systems are not yet prepared to handle a sudden incoming light surge of that nature, and it might result in some loss of troops.

Mission Control does not wish to outline the emergencies that would cause us to activate all members of the off- and on-planet units prematurely. We do not want your attention on these matters, as your focus may cause them to occur. Therefore, it must be sufficient for you to simply know that there are emergency crews on stand-by. They are prepared to assist the planet should the birthing process become too difficult at any point. As previously stated, we are here to assure this planet's safe delivery into the light. Any necessary measure is within our jurisdiction and capacity. That is all we care to say on this issue at this time.

Mishaps of the Mission

This is another topic that Mission Control does not wish to dwell on because attention on the casualty list will only serve to increase it. All we will say on this matter is that all missions to dysfunctional planets have their dangers. Some members of the mission have effectively crash-landed and may not recover sufficiently in time to complete their assignments. Others have become so embroiled in their dysfunctionality that they have completely forgotten the point of assuming the condition in the first place. There is still time to rectify some of these mishaps; however, a few mission members are in serious shape and their prognosis isn't very good.

Although only a very minor fraction of our ground crew is in trouble, we would like these members to know that the injuries they have sustained are not an indication of failure in our eyes. We regret that there are any casualties, but it is impossible to assure total safety to all mission participants because such assurance would be a violation of the free will and the divine sovereignty of our mission members. Mission Control would like our wounded members to know that their efforts have been deeply appreciated and much more than Purple Hearts awaits them when they return home.

5

Assistance

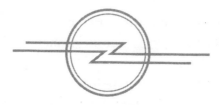

THE FUNCTION of this section of the manual is to help you remember the vastness of the community you come from and the loving assistance the members of that community willingly extend to you all. We know it can feel lonely in the spiritual fast lane on this planet. But remember that this is only a feeling and not a fact. Please avail yourselves of the resources that are your birthright and know how much you are loved. This is Mission Control, on behalf of all the great Forces of Light, completely at your service.

Foreign Aid

Throughout the process of your awakening and beyond, you have the right to call upon Mission Control, the Council Seats, all Alliances, Federations, and Confederacies of Light, the Ascended Masters, and every kingdom of this planet for assistance. That is just to name a few of your resources. You are a member in good standing of this mission and we urge you to exercise your rights. This is not only for your protection; it is the very means by which higher dimensional energies can legitimately gain entry onto this plane.

As we have stated before, *we do not invade*. However, when a member of this mission has earned the status of Earthling by living it and then requests interdimensional intervention or support, we can legitimately answer that request without violating Universal Law. In this manner, our presence will continue to infiltrate this planet for the purpose of bringing it into alignment with the greater Universe of Light. We hope this information makes it clear that "E.T., phone home" is actually very sound advice.

Morale Boosting

Mission Control is aware that many of you are very weary and discouraged with this mission. From your current perception, we know it looks pretty bleak down there. To watch world systems in decay, ambulatory insanity at the helm, denial on a rampage, humanity down each other's throats, and a dying biosphere is probably not your idea of a good time.

Try to remember that though you may be vastly outnumbered on this planet, you belong to a greater family that is by far in the majority of the remainder of the universe. Align with your heritage, remember your birthright, and be certain of your destiny. You are children of the stars, sired by light, and your reality is the superior one. The damage and corruption you see around you is just the ending choreography of the old world's last dance, and the promised reclamation of this planet is but the final manifestation of a campaign that has already been won.

Curing Battle Fatigue

The best cure for battle fatigue is not to battle. Although you may find it difficult not to inject a little sanity into the lemminglike rush toward death you see all around you, do *not* intercede. The old world is dying. It must and will come down. The best you can do is allow it to die as gracefully as possible.

Whatever you put your attention on increases. For the sake of the ecosystem and the new emerging civilization, remove your attention from the death process and place it on the process of birth instead. Misplaced attention will just act to prolong the ending's agony and delay your inevitable, exalted future.

The Special Forces

Because we do not have a millennium to spare, Mission Control has not left the process of reawakening solely in your hands. Alliances, commands, and transition teams have been sent in to facilitate your awakening and help snap you out of your coma. Please be on the lookout for these energies.

You will be able to identify the Special Forces primarily by your inner response to them which was pre-encoded into your DNA structure before you left. No matter how "rational" you believe yourself to be, you will find yourself strangely interested in the unbelievable things they are saying without knowing why.

The Special Forces are distinguishable from Earth-based organizations in that they do not lie, are not wimpy, and don't want followers. They will not allow you to use them to replace worn-out, fear-based, disempowering religious belief systems. They will insist on your sovereignty, refuse to be outside authorities, and will not allow you to dump your responsibility or power at their feet. Their purpose is clear and simple: They are here to assist you into your full presence so they can then aid in co-creating a new reality with their peers.

Another characteristic feature of the Special Forces is a well developed sense of humor, also distinguishing them from most Earth-based "spiritual" groups. These Forces may be facilitated by walk-ins. They may use art forms, such as dance, interdimensional languages translated into tones, or whatever else they can get their hands on to circumvent your linear, two-dimensional linguistic systems. They are experts in the transmutative process and use other dimensional technologies to break through dysfunctional patterning.

Mission Control's primary goal is to successfully complete this mission with as little loss as possible. Please do not ignore the Special Forces that were sent in for your benefit. They are the Green Berets of this mission.

Interdimensional Brain Surgery

Do not be alarmed by the subject of this article. The only dimension on which brain surgery is dangerous is the third. Every other dimension (not counting the first, second, and fourth) has it down pat, and malpractice suits are virtually unheard of. Interdimensional brain surgery is another form of assistance we offer you.

This surgical procedure enables us to reroute dysfunctional brain patterns, rewire circuitry that has shorted due to deranged thought-form overloads, cure all computer viruses that your brain may have contracted, and replace existing fuses with heavier equipment to insure that everything doesn't blow out when all the lights come on.

To operate, however, we need your permission on one level or another—conscious permission preferred. Even your medical profession has gotten that far, usually having you sign a release before they nearly or actually kill you. The difference in our request is that it is not motivated by a desire to stay out of court but by our total respect for your sovereignty.

For those of you who are reluctant on any level to give your permission to go under the knife, you may be relieved to know that we don't use knives. It may also be helpful to know that we haven't lost a patient yet. Mission Control awaits your decision.

Exploratory Emotional Body Surgery

Unlike interdimensional brain surgery, you do have reason to be alarmed by the subject of this article. In answer to your question, "Will it hurt?" the answer is "Yes." This surgical procedure requires conscious participation and cannot be done under anaesthetics. In fact, many of you will have to come out of the anaesthetics you are currently under in order to participate.

If you enjoy going where no man or woman has gone before and are not put off by a sloggy journey through your own internal swamp, this surgery will present little or no problem. However, if you are squeamish about traveling over darkened and repulsive terrain, we suggest you toughen up, because there is no way around this one. Lightness and darkness cannot coexist in the same place at the same time.

Although emotional surgery requires some bravery, Mission Control would like to remind you that no one in their right mind would have signed up for this particular mission if they did not have any courage. The fact is, the only thing more painful than going through this procedure is *not* going through it. Our surgical staff is at your disposal and ready to assist you through this process.

Creative License

If you do not already have a Creative License, we suggest you apply for one immediately. We assure you it will come in very handy as you try to accomplish what it is you came here to do.

When you send in your application, be sure to indicate the level of creation you feel you are capable of handling. Once your application is received, Mission Control will check its own files to see if the class of license you have applied for matches our data concerning the creative skills you can manage. Even though Mission Control already knows the answer, we ask for your self appraisal just to check your understanding of your role in the co-creative process.

In most cases, the class of license you request will be well below the level you can handle, in which event you will be issued a Learner's Permit. Please do not be insulted if this is what you receive. It is temporary and will be replaced by your real license as soon as you fully awaken to your creative capacity. The Learner's Permit is simply a safety precaution. A full-fledged Creative License requires total conscious control of the reality you are designing. It also grants you "driving" privileges outside your dimension. Unfortunately, losing control of your vehicle interdimensionally can cause an even worse traffic hazard than it does within the relatively safe confines of your planet.

Although you will not be asked to take a written exam, a heart/mind coordination test is a must. This mandatory examination will be administered to you on another dimension by our DMV staff. Also, when applying for your Creative License, you need not indicate whether you wear glasses, contacts, or are legally blind. Just tell us if you can see; that's all we want to know.

⚡

Note: If you wish to apply for a license, please refer to the ET Census Form at the back of this book for our address.

Recent Legislation

Since time is almost over (and without it, it's impossible to live out your lives on the old "go now, pay later" plan), the Stellar Councils have unanimously voted to repeal the Law of Karma. This came about because Mission Control brought it to the Councils' attention that there wasn't enough physical time left to fulfill the Law of Karma's requirements and still meet our transmutative deadline. As a result, the Councils decided that it was easier to get rid of the whole thing than it was to figure out a way to meet its demands. Another reason the Councils were moved to this decision is that the Akashic Records are just about full. The thought of having to add on another wing and increase its library staff was more than the participating Councils cared to address at this time. They felt they had more pressing projects to invest in during this fiscal millennium.

As a result of these factors and the additional fact that it is virtually impossible for anyone to be a master and a student at the same time, the Councils have not only rescinded the law governing the karmic educational system, but have also unanimously voted to enact the Law of Grace. Consequently, all debits have been removed from the cosmic records and you are free to move forward with no reference to any debts you may have incurred. You are also free to stop pretending that you are a student. This legislation makes it easier all around and has sent a sigh of relief throughout the Intergalactic Council's administrative staff—especially the Justice Department and the Interdimensional Retribution Service. It should likewise send a sigh of relief through you.

Mission Control repeats this important bit of news: The Law of Karma has been repealed and the Law of Grace enacted to assist you in your manifestation of divinity. All debts have been forgiven and all court dates canceled. You are free to proceed outside the jurisdiction of karma and in the state of grace. The blessings of all the Councils go with you.

Self-help Techniques

The greatest self-help technique you can practice is the art of laughing. This is not to say that everything that is coming down is entirely funny. For instance, you may find it difficult, at first, to get a chuckle out of a rapidly disappearing ozone layer and the petrochemicals you are drinking with every glass of water. And, to be perfectly honest, even *we* don't find the Federal Reserve very amusing. However, getting depressed is not an answer.

This is the most critical moment of change in this planet's history, and your assistance in that change is vital. Humor has the effect of raising your vibratory level, and you won't believe how high it has got to go to get through this one. Going catatonic over the seriousness of the global situation will not only not help the globe, but it will also effectively knock you out of the ballgame. Our advice? Keep laughing.

Another practice you will find invaluable is owning up to your creative capacity. Your reality is formed by your attention, and it is entirely your choice if you end up as a second-rate actor in a B movie instead of a star on the star that is about to be born. It is also advisable to keep in mind that you are here on assignment. Please don't get sidetracked into thinking you have cancer just because you have visited the ward. *Remember who you are and what you are doing and keep your eyes on the stars.*

The Great Awakening

The 1990s are the decade of The Great Awakening. By comparison, the '90s are destined to make the '60s look like little more than an episode out of *Leave It to Beaver.* In this decade, the second wave of extraterrestrials will remember who they are.

This newly awakening group constitutes the majority of the beings on this planet who are carrying within their genetic structure the seeds of a new consciousness. This tide of consciousness is an unstoppable force, and its impact is destined to sweep across and shape the shores of the incoming millennium. The Great Awakening is a manifestation of the Victory of Light that has already been accomplished beyond this plane and now has only to play itself out on this dimension.

The greatest help you will receive on this mission will happen through this awakening of your fellow members. The escalation of transmutational energy caused by this awakening will irreversibly tip the global scales in the direction of spiritual realignment. This vibrational escalation will be a demonstration of a very sophisticated, extraterrestrial concept which we call multidimensional marketing. During these times, please be generous and loving in your assistance to those around you. They are most likely your down-line.

Starseed—The Next Generation

Another great source of assistance on this mission will be extended to you by the generation that follows. This manual is primarily directed at the van-

guard of this mission whose task is to cut the pathway to a new civilization. However, the generation
that you have prepared the way for is right behind
you. They are the builders of the civilization for
which you now establish the foundation.

We have noticed that your current civilization has
been alarmed by this generation, as they have begun
to make mincemeat out of your standardized tests of
measurement. Many of them are logging remarkably
low scores on your intelligence tests, such as the SAT
examination. They are also having a field day with
your psychological tests for normalcy, such as the
Minnesota Multiphasic Personality Inventory. May
we suggest that the Minnesota Multiphasic has never
been adequate to measure anyone outside of Minnesota, and it is even more inadequate in measuring
an extraterrestrial who may have an alarming predisposition toward androgyny and other psychologically
suspect behavior. This starseed group is equipped
differently and is basically bored by the questionable
standards of intelligence and dysfunctionality posing
as mental health that you are submitting them to. It
would actually be more appropriate to measure this
generation with a Richter scale if you were truly interested in understanding who they are. And it may
be time to scrap the outdated exams that only assess
their response to slavery.

Just as you are great masters of consciousness, so are they. Their task is slightly different, but they will support you in yours as they await their moment to make their presence known. Treat these masters well. They are the seeds that are to bear the fruit of your ecstatic destiny.

Audio-visual Aids

The Intergalactic Council is in the process of considering its forthcoming line of interdimensional paraphernalia to assist you into your real identity. If you are already there, none of these audio-visual aids will be necessary. However, if you are still in transition, you may find their "fall line" useful. If you wish to be on the Council's mailing list, please send your name and address to us at the address listed in the back of this book. (Be sure to indicate that you are interested in the Council's E.T. Designer Line so that we don't mistakenly issue you an unrequested permit or license.)

⚡

Note: If you have applied for a Creative License or have filled out your address and sent in our census information at the back of the manual, you will automatically be placed on this mailing list unless you indicate otherwise.

Monitoring

This entry is not for your assistance; it is for ours. Some of our technicians have lodged a complaint and requested that we place it in this manual. As mentioned elsewhere, all members of this mission are under constant surveillance by our monitoring staff. In many cases, this has gotten to be quite a bore, and some of our personnel are having trouble staying awake at their panels. They are wondering whether you have forgotten why you are on this planet and would appreciate a little more activity in conjunction with the mission. So, for their sake as well as the planet's, will you please step on it? Their job description does not include monitoring an entire squadron that is asleep at the wheel.

A FORMAL INVITATION

Mission Control
respectfully
requests
your presence
at
a come-as-you-are party
RSVP

WE KNOW that this mission is not easy. We also know that many tears have been shed in the awesome process of its spiritual unfoldment. Be consoled in knowing that all tears are soon to be wiped from your eyes and all your pain dismissed and forgotten. The glory and joy of what is about to transpire will render all you have endured a minor expense, a price you would be more than willing to pay again.

You are all cordially invited to attend the birthday celebration that marks this mission's successful end. This celebration will put all the combined feasts of every earthly head of state to shame. Such feasts will seem pathetic gestures compared to the party that Spirit is about to throw in your name. The revelation of the nature of your presence on this plane will soon be announced. You will be known as the honored guests of the Spirit you came to serve, a disclosure that will lend an entirely new meaning to the saying "a star-studded cast."

The dress code is simple but mandatory. You must come clothed in your full presence, dressed in the spiritual light of the Lords that you are. Clothe yourselves in the finery that befits Spirit's messengers to this plane. Come out of your hiding and come fully attired as the distinguished members of Spirit's divine delivery system of the stars. Come, in short, as you truly are.

We bless you all for your courage and your commitment and we honor you for your accomplishments on this plane. Take heart in the knowledge that your task is almost over. This carbon-based planet will shortly burst into a diamond, a gem in the crown of this solar system's skies. The celebration will then begin.

This is Mission Control
Over and out

THE EXTRATERRESTRIAL CENSUS

Because the number of entities that comprise this mission is vast, and many have entered this plane representing a myriad of alliances, commands, councils, and federations, we have decided to run a census. Although we usually run a census every other millennium whether we need it or not, this particular census is of special significance to us. That is because, due to the mission, there are many more extraterrestrials here than usual. There is also some importance attached to this census because the Council's statistics have become confused by the fact that many mission members are part of group soul endeavors which have splintered onto this plane for the occasion. Our records count these members as one. (The fact is, that is how we count the entire mission, no matter which planet, galaxy, or dimension you may be from or which alliance or federation you are affiliated with.)

Counting the whole thing as one, or a group soul project as one, is fine on the fifth dimension but is causing us a bit of a traffic jam on the third, where some of our "ones" are now numbering over five million. Just to straighten this matter out, we are asking that you voluntarily turn yourselves in. Although our statistics will never add up to a count above one, it is a matter of both interest and tidiness to know where our mission fragments are located and how they think they are doing.

Beyond satisfying these superficial third-dimensional concerns, this census is actually being taken for your benefit. Your incoming data will be used to energetically bind you in your awareness of each other on this plane. We, in truth, don't need a count to know who you are, where you are, and what you're up to. (Put another way: We know if you've been sleeping. We know if you're awake. We know if you've been bad or good, etc.)

—*The Intergalactic Council*

OFFICIAL EXTRATERRESTRIAL CENSUS FORM

Another two-thousand-year cycle is winding down, and it is once again time for our bimillennial extraterrestrial census. As usual, the Intergalactic Council guarantees that your census form remains confidential. Until the end of time, no one but members of the Census Bureau staff, Intergalactic Council members and the Council's various departments' personnel, affiliate members of the mission including all Alliances, Commands, Federations, etc., interrelated Stellar or Interdimensional Council delegates, the Commander of the Royal Celestial Air Force, its officers and crew, Mission Control, and the members of these participants' immediate families will be permitted to see your form. Thank you for taking time to complete and return this questionnaire. It's important to you, your community, your nation, and the Planet.

Note: The following census form has to be either torn out or copied, trimmed, filled in, folded, and stuffed in an envelope before mailing. This is a test of your manual dexterity and your attention span. You must then purchased first-class postage and put it on the envelope by yourself. This is a test of your interest level. If this process seems like too much trouble, don't forget we have watched you go to more elaborate efforts in your attempts to become a millionaire by entering magazine contests. Surely you can take a moment out for this co-creative venture designed to help liberate you from ever having to enter a third-dimensional sweepstakes again.

1a. I am: ☐ An Extraterrestrial Master

☐ A member of the Angelic Hosts

☐ A Council, Federation, Alliance and/or Command member

☐ An Interdimensional Adept, Master, or Lord

☐ A member of the Special Forces

☐ A Group Soul posing as many human beings

☐ All of the above

☐ All of the above except_____

☐ Other (please specify)_____

1b. My location is: Longitude_____ Latitude _____

(if unknown, or if you wish to be on our mailing list, please fill out address)

Name _____

Street _____

City_____ State_____ Zip Code_____

Country_____ Phone (____) _____

2. Are any other members of your household also extraterrestrials?
☐ Yes ☐ No If your answer was yes, how many? _____

3. Sex: ☐ Male ☐ Female ☐ Androgyn ☐ Uncommitted

4. Race: ☐ Human ☐ Other (If you are some form of android or if you insist on remaining an alien, please do not answer this question. That is not what we mean by "other.")

5. I am a (check one): ☐ Vanguard E.T. ☐ 2nd wave E.T. ☐ Second generation E.T. ☐ Nouveau E.T.
My age is _____ (This information is solely to see if you are anywhere even near the ballpark in correctly identifying your position.)

6. The following question is designed to determine your mental attitude toward the mission. We have absolutely no intention of altering the mission based on your responses, so feel free to answer in whatever manner you wish. Not only can we take it—we've heard it all.

My honest opinion about/burning question concerning/personal denial regarding this mission is:

☐ I don't know what you're talking about. What mission?

☐ Feel free to beam me up. There really *isn't* any intelligent life down here.

☐ This planet is in no shape for an evolutionary leap. It couldn't even withstand an evolutionary hop, and I advise immediate reconsideration of the entire plan.

☐ I don't remember signing up for this mission. If I did, is there any such thing as a discharge? Dishonorable is fine with me.

☐ I believe I have struck upon a better planetary transition plan which does not require my direct participation. Can we talk?

☐ I live in Wilton, Connecticut, and have mistaken my portfolio for my identity. Is there any possibility of me and my portfolio being restationed at this point?

☐ This mission is a piece of cake. Where to next, big guys? (Be advised that if you check this box, further questioning may be necessary.)

☐ My personal and as yet unexpressed thoughts on this matter are: (25 words or less, please) _____

_____

_____

7. I expect to be fully able to report for active, conscious duty no later than: Month ____ Year ____ (Anyone speculating on a year later than 1998 and who is over the age of twelve, please check here to request an extension. ☐)

Thank you for your cooperation.

Please send your completed census form to the following address:

The Intergalactic Council • Bureau of the Census •
P.O. Box 2066 • Pagosa Springs, Colorado 81147 • USA

(If you wish further information about our activities on this planet, please contact our third-dimensional offices for an update. If you would like a Creative License, please include a one-time fee of $11 in US Earth funds only.)